The Joiner's Journey

A Detailed Exploration of Fine Woodworking

Clara Hartley

Contents

1 Introduction to Joinery

Choosing the right wood is paramount. Consider the project's intended use; a sturdy table demands robust hardwoods like oak or maple, while a delicate jewelry box might suit softer woods like cherry or walnut. Examine the wood grain closely—tight, straight grains indicate strength and stability, while wild grain can be more challenging to work with but can also add unique character. Be sure to account for wood movement due to changes in humidity; selecting appropriately seasoned wood minimizes warping and cracking. Understanding the differences between hardwoods and softwoods is fundamental. Hardwoods come from deciduous trees and are generally denser and more durable. Softwoods, from conifers, are often easier to work with but less resistant to wear. Remember, even within species, variations exist in density and grain patterns. Finally, inspect each piece meticulously for knots, cracks, or other imperfections that could compromise its structural integrity or affect the final aesthetic appeal.

Proper preparation prevents poor performance. Begin by milling your wood to the precise dimensions required for your project. This involves using a planer and jointer to achieve perfectly flat and straight surfaces—critical for creating tight-fitting joints. If your wood is not already seasoned, allow ample time for it to acclimate to your workshop's environment before working with it. This prevents sudden changes in moisture content that can cause warping or cracking. Paying attention to details at this stage will save you frustration later. Proper preparation isn't just about milling; it also includes carefully selecting your wood and checking for defects. Remember that you can also use joinery to hide imperfections in your wood if they are not structural.

Understanding basic joinery tools is crucial for success. A sharp chisel, essential for creating mortises and cleaning up joints, requires regular honing and sharpening. Always use appropriate safety precautions while using sharp tools; keeping your work area clean and organized minimizes the risk of accidents. A mallet, carefully chosen for its weight and balance, is a crucial companion to your chisels, enabling precise blows for creating joints. Similarly, your hand saw should be sharp and properly set to allow for accurate cuts, and clamps, ranging in size and style, are vital for holding workpieces securely in place during assembly. A well-equipped joiner's bench allows for accurate work and reduces fatigue. Proper care and maintenance for your tools will enhance your work and safety. Remember that having the right tools makes the project much easier and less stressful. Don't be afraid to invest in quality tools that will last. Safety should always be your primary concern.

Layout and marking are where accuracy begins. Using a sharp pencil and a steel rule, carefully mark the layout of your joints onto the wood. Double-check your measurements to ensure precision, avoiding costly errors further down the line. A marking gauge proves invaluable for creating consistent lines, particularly for repetitive tasks such as creating mortises or laying out dovetails. Always practice good woodworking techniques and safety practices. A combination square aids in achieving precise 90-degree angles, while a bevel gauge helps create angled cuts with accuracy. Having the right marking tools and techniques will lead to consistently accurate results, making your project stronger and more aesthetically pleasing. Precision is key in joinery; taking the time to do this properly will save you significant time and effort in the long run.

Mastering basic joints is a foundational skill. The simple butt joint, while seemingly basic, teaches the importance of accurate cuts and glue application for a strong bond. It's a simple first step on the journey of joinery. The lap joint provides superior strength, its overlapping surfaces creating a sturdy connection and, with practice, you will learn that simple adjustments make the difference between a quick, easily made joint and one that's highly secure and visually appealing. Consider the half-lap or the miter joint, each offering unique solutions for particular applications. The versatility of these elementary joints lies in their adaptable nature, allowing for creative solutions in diverse woodworking projects. The more you practice, the more confident you will become in your ability to execute these joints precisely and efficiently.

Remember, consistent practice is key. Begin with simple projects, gradually increasing complexity as your skills develop. Don't be discouraged by initial challenges; every skilled joiner has faced setbacks. Embrace these experiences as opportunities for learning and growth. Experimentation is crucial; try different techniques and approaches to find what works best for you. Take advantage of online resources, books, and workshops to enhance your knowledge. Continuous learning is important to stay updated on the best practices and techniques. Engage with the community, sharing experiences and receiving feedback, fosters a better understanding and improves skills. Joinery is a journey, not a destination; embrace the process, celebrate the small victories, and enjoy the craft.

1.1 Basic Tools & Safety

A well-equipped workshop is your foundation. Start with the basics: a hand saw for accurate cuts, a smoothing plane to level surfaces, chisels for precise shaping, and a mallet for controlled blows. Don't overlook clamps; they're essential for holding pieces together securely during gluing. A good quality measuring tape and combination square are crucial for accuracy; invest in tools that feel good in your hand, improving your precision and reducing fatigue. Consider a workbench; it provides a stable, dedicated work surface, and protects your work from damage. Protecting yourself is just as important. Safety glasses are non-negotiable; even small wood splinters can cause eye injuries. A dust mask is crucial for preventing inhalation of wood dust, a known respiratory irritant. Ear protection is vital, particularly when using power tools. Always unplug power tools before making adjustments or cleaning them. Learn the proper techniques for each tool, and remember, rushing leads to mistakes—and potentially injuries. Proper posture is vital too; your back will thank you later.

Choosing the right wood for your project is half the battle. Different woods possess unique properties. Hardwoods like oak and maple are durable and ideal for furniture, while softer woods such as pine are easier to work with for beginners. Consider the wood's grain pattern, density, and overall appearance. Before starting any project, inspect your wood carefully for knots, cracks, or other imperfections. These imperfections can weaken your piece and affect the final aesthetic result. Proper wood preparation is essential for successful joinery. Start by milling the wood to the desired dimensions, using a planer or hand tools for smooth, accurate surfaces. If the wood is too wet, it will shrink and potentially crack, creating a poor foundation for joinery. If it's too dry, it can split. Understanding wood moisture content is vital. Proper conditioning ensures dimensional stability of your project. A moisture meter is a valuable tool for determining the moisture content of your wood. Always store your wood correctly to prevent warping, cracking or mold, and protect your investment.

Sharpening your tools is crucial. Dull tools are inefficient, require more effort, and increase your risk of injury. Learn to sharpen your chisels, planes, and saws regularly, using a sharpening stone or honing guide. A sharp tool effortlessly slices through wood, resulting in clean, precise cuts. This is especially true when working with hardwoods. Regular maintenance keeps your tools performing at their best; it's an investment in the quality of your work and safety. Learn different sharpening techniques according to your tools. Using the right method for your tools maximizes efficiency and prolongs tool lifespan. Remember to use appropriate lubricants while sharpening. The goal is not only sharp tools, but also well-maintained ones, for efficient and safe woodworking.

Working with wood involves inherent risks, so understanding and implementing safety precautions is paramount. Always be mindful of your surroundings; avoid distractions and maintain a clean, organized workspace. Keep your tools properly stored, out of reach of children.

Know the properties of the wood you're working with; some woods are harder, and therefore more likely to cause splinters. Always use appropriate safety equipment, such as eye protection, hearing protection, and a dust mask. Never operate power tools if you're tired or feeling unwell. If you have any doubts, take a break. Safety is not a suggestion; it's a non-negotiable requirement for successful woodworking. Proper lighting in your workspace also promotes safety; reduce shadows that could lead to accidents. Remember that the proper use and maintenance of your tools are vital to both their longevity and to your personal safety.

Remember, patience and practice are key to mastering woodworking. Don't get discouraged if your first few projects aren't perfect. Each project is a learning experience. Take your time, focus on accuracy, and enjoy the process. Fine woodworking is a craft that rewards dedication and patience. The more you practice and the more projects you complete, the more proficient you'll become, and the more confident you'll feel in your abilities. Enjoy the journey. The satisfaction of creating something beautiful and functional with your own hands is incredibly rewarding. Start small, gradually increasing the complexity of your projects as your skills improve. There's a whole world of woodworking projects waiting to be discovered! Start simple, and gradually expand your repertoire of tools and techniques as your skills develop. This gradual approach will build your confidence and ensure a smoother learning experience. Never underestimate the power of practice.

1.2 Wood Selection & Preparation

Choosing the right wood is paramount. Consider the project's demands – a sturdy table needs strong, durable wood, while a delicate box might suit a softer, more easily worked species. Hardwoods like oak, maple, and walnut offer strength and beauty, perfect for furniture. Softer woods such as pine or basswood are easier to carve and work with, ideal for beginners and projects requiring intricate details. Research different species to understand their unique properties; grain patterns, density, and workability all affect the final outcome. Always source wood from reputable suppliers ensuring responsible forestry practices. Knowing the wood's origin and how it was harvested will inform your understanding of its qualities, and support sustainable practices.

Proper preparation starts with milling. This involves cutting the wood to the correct dimensions, ensuring straight, square edges. A table saw, jointer, and planer are invaluable tools for achieving this precision. Always use safety equipment, including hearing protection and dust masks, throughout the process. Precise cuts minimize waste and enhance the structural integrity of your final product, which is essential for long-lasting, high-quality joinery. Careful milling also ensures all pieces will fit together flawlessly. Remember that accuracy here prevents significant problems down the line.

Drying is crucial. Freshly cut wood contains significant moisture. Improper drying can lead to warping, cracking, and movement. Air drying is a slow, natural process, suitable for smaller projects or if you have the patience. Kiln drying is faster, using controlled temperatures and humidity to remove moisture efficiently. For high-precision joinery, kiln-dried wood is often preferred for its dimensional stability. Regardless of the method, monitor the moisture content using a moisture meter. The ideal moisture content for indoor woodworking is typically around 6-8%. Understanding the drying process, along with moisture content monitoring, is vital in preventing future problems.

Before starting any joinery, the wood needs conditioning. This involves acclimating the wood to the environment where the finished piece will reside. Sudden changes in humidity and temperature can cause movement, potentially ruining your work. Allow the wood to sit in your workshop for several weeks, allowing it to adjust to the ambient conditions. This step is often overlooked but is exceptionally important for the long-term stability and integrity of your finished product. Taking the time to properly condition your wood will drastically improve the longevity of your woodworking projects.

Wood defects can be a significant challenge. Knots, splits, and checks are common imperfections that can weaken the wood. Carefully inspect your wood boards before milling, identifying and avoiding these defects whenever possible. Sometimes, creative design adjustments can incorporate minor flaws, adding character to the finished piece. Severe defects, however, might necessitate discarding the board. Careful inspection and selection prevent problems during construction, saving time and materials in the long run, and increasing the likelihood of a quality result. Selecting high-quality lumber should always be prioritized.

Finally, remember that preparing wood is an iterative process. Each project presents unique challenges. The more experience you gain, the better you'll become at selecting and preparing wood for your projects. Don't be afraid to experiment, and always learn from your mistakes. Each project should be a learning opportunity. Over time, you'll develop an intuition for choosing and preparing the perfect wood for any project, ultimately leading to higher-quality outcomes. The preparation stage is fundamental; neglecting it leads to significant problems in later stages.

2 Essential Joints

The mortise and tenon joint, a cornerstone of woodworking, deserves meticulous attention. Its strength lies in the precisely sized mortise—the hole—and tenon—the projecting piece—which interlock, creating a strong and durable connection. Begin by accurately marking your mortise location on the workpiece. Use a chisel and mallet for a traditional approach, meticulously removing wood to create a clean, square hole. Alternatively, a mortising machine offers speed and precision, especially for repetitive tasks. Remember, accuracy is paramount; a poorly cut mortise will compromise the joint's strength and appearance. The tenon, the mating piece, requires equally careful attention. It should be precisely sized to fit snugly within the mortise, neither too loose nor too tight. Test the fit frequently as you shape the tenon using chisels or a tenon saw. A well-executed mortise and tenon joint is almost invisible, a testament to the joiner's skill and precision. Consider using glue to further enhance the joint's strength and longevity; it adds a crucial layer of security that enhances the overall integrity of the construction.

Dovetail joints, known for their elegance and strength, represent a significant step up in complexity. Through-dovetails, where the tails and pins completely penetrate the adjoining pieces, offer exceptional strength—they are frequently used in drawers and boxes. However, mastering this joint demands patience and practice. Accurately marking the layout is critical, requiring precise measurement and careful attention to detail. A marking gauge will help you achieve consistency. Then, carefully cut the tails on one piece and the pins on the other, using a sharp chisel and mallet. Each cut must be clean and precise; sloppy cuts weaken the joint and compromise its appearance. Test the fit frequently, adjusting as needed, before finally assembling the joint with wood glue. Half-blind dovetails, where the tails are fully visible but the pins are partially hidden, offer a cleaner aesthetic for applications where the joint is more exposed. They are more challenging, technically, and mastering them truly exhibits craftsmanship. Remember the importance of practicing this type of joinery on scrap wood before moving on to your project.

Box joints, a simpler alternative to dovetails, offer exceptional strength and are often preferred for boxes and drawers where aesthetics are less critical. These joints are characterized by their series of uniformly sized pins and mortises that interlock, creating a remarkably stable connection. Laying out the joints requires careful attention, just as in the dovetail. Accurate measurements are essential to ensure the consistent sizing of the pins and mortises. Using a router and jig is highly efficient for cutting these joints. This allows for speed and accuracy, particularly useful for a large number of identical joints, ensuring consistent results. For those without a router, using a hand saw and chisel is also perfectly viable—it's just more time-consuming. The key is clean cuts and a snug fit. Once assembled, the joint's strength is substantial—it's surprising how durable these joints can be for their relative simplicity, which is why they're a favorite for many furniture makers. Ensure that the boxes are square and the joints are flush, contributing to the overall professionalism of the final product.

Beyond these essential joints, many variations and adaptations exist. Learning to adapt these fundamental techniques—modifying them to suit different situations—demonstrates a deeper understanding of joinery and woodworking principles. The strength and elegance of a finished piece frequently depend on the selection of the proper joint, a consideration often neglected by beginner craftspeople. It is essential to assess the needs of the project; some joints offer superior strength while others are aesthetically preferable. Experimentation and practice are invaluable in expanding your skills and developing a refined approach to joinery. The subtle nuances of each method can easily translate to noticeable differences in the outcome. As your proficiency increases, you'll find yourself intuitively selecting the most appropriate joint for each application, a skill that marks the transition from beginner to confident craftsperson. Remember, the most beautiful and durable pieces are often built upon the foundation of simple, flawlessly executed joints. Never underestimate the impact of preparation and careful planning. The most sophisticated designs often start with these basic building blocks.

2.1 Mortise and Tenon

The mortise and tenon joint, a cornerstone of woodworking, is characterized by a projecting tenon (the "tongue") fitting snugly into a mortise (the "hole") in another piece of wood. Its simplicity belies its strength and versatility, making it ideal for a wide range of applications, from chair legs to table frames. Creating a strong and accurate mortise and tenon requires careful planning and precise execution. Starting with well-seasoned lumber, free from knots or imperfections in the areas to be joined, is paramount. This minimizes the risk of splitting during the cutting process and ensures the longevity of your finished piece. A properly prepared workpiece is the foundation of a successful project.

Accurate marking is absolutely crucial. Use a marking gauge to precisely determine the tenon width and shoulder lines. A sharp chisel, paired with a mallet, is essential for creating clean, straight mortises.

Begin by carefully paring away the waste wood, working methodically from the edges and gradually deepening the mortise. Consistent, gentle blows are key; avoid forceful strikes that might damage the surrounding wood or cause the chisel to slip. Regularly check your progress using a try square to ensure the mortise is perfectly square and the correct depth. Precision at this stage translates directly to the strength and accuracy of the final joint.

The tenon itself needs to be equally precise. After marking the tenon shoulders and cheeks, use a tenon saw to carefully cut the cheeks to the correct depth. Again, precision is vital. A slight inaccuracy here will significantly compromise the fit of the joint. The shoulder cuts should be perfectly square, and the cheeks should be smooth and parallel to one another. Many woodworkers prefer using a shoulder plane to refine the tenon shoulders, ensuring a flush fit with the mating piece of wood. This step allows for an excellent surface-to-surface connection, which is crucial for strength.

Once both mortise and tenon are cut, test the fit. It should be tight but not overly so, allowing the tenon to slide smoothly into the mortise. A slightly loose fit can be corrected with the use of a plane, carefully paring away small amounts of wood until the desired snugness is achieved. However, a tenon too tight will risk splitting the wood. This final fitting stage provides an opportunity to fine tune and achieve perfection. Only after this refinement will you glue your creation. The use of wood glue and clamps will be sufficient to bond this joint successfully.

The type of wood influences your approach. Hardwood, like oak or maple, will demand sharper tools and more careful execution, while softer woods like pine may be easier to work with. The specific dimensions of the mortise and tenon will also vary depending on the project, influenced by the size and weight of the pieces being joined. However, a general rule is that larger joinery generally necessitates thicker joinery. Consider the overall design of your project—a visually appealing joint should be as functional as it is beautiful.

Remember, the true mastery of the mortise and tenon lies not only in the precision of the cut but in the understanding of the wood itself. Pay attention to the grain direction, ensuring that your cuts align with the wood fibers to minimize the risk of splitting. Understanding the properties of the wood you are using is fundamental to success. Wood is a natural material, and its variations in density, grain, and moisture content will all influence the cutting process. Practice makes perfect, and repetition will enhance your skills and speed. Begin with smaller projects and gradually increase the complexity.

Ultimately, the mortise and tenon is more than just a joint; it's a testament to the skill and precision of the woodworker. The satisfying click as the tenon finds its home in the mortise is a moment of quiet achievement, a reward for careful planning and precise execution. Mastering this fundamental joint is a crucial step in any woodworker's journey. With careful preparation and precise technique, it will become second nature. The enduring strength and pleasing aesthetic of the mortise and tenon joint ensures its continued relevance in the world of woodworking. This will provide the necessary foundation to proceed to more advanced woodworking projects.

2.2 Dovetail Joints

The strength of a dovetail joint lies in its interlocking nature, a clever design that creates a remarkably strong bond. Unlike other joints, the dovetail's strength isn't solely reliant on glue; the interlocking shape provides significant mechanical integrity, making it a favorite among fine woodworkers for its durability and visual appeal. This self-aligning aspect is key; each tail precisely fits into its corresponding pin, minimizing the need for excessive force during assembly. The result is a robust joint that can withstand significant stress, far exceeding the capabilities of simpler joinery methods.

Begin with the careful layout of your tails and pins. Accuracy is paramount here. Using a marking gauge, precisely measure and mark the spacing for your tails on the end grain of the workpiece. Ensure consistent spacing for a visually appealing and structurally sound joint. The tails should be consistently spaced and sized, and the pins should perfectly mirror those markings. An error here can lead to problems later and compromise the structural integrity of the entire piece. Consider using a template to ensure consistency, especially when working with multiple joints. Precision is a cornerstone of fine woodworking, and meticulous layout is the foundation of a perfect dovetail.

Cutting the tails requires a sharp chisel and a steady hand. Begin by carefully paring away the wood between the marked lines, working from the waste side to avoid damaging your carefully laid-out lines. Use your chisel to create clean, accurately-angled cuts, ensuring that each tail is the same width. Patience and a light touch are crucial to avoid mistakes. Remember, you are shaping wood, not destroying it; each cut should remove only a small amount of material. Regular sharpening of your chisel will make this process easier and more accurate, resulting in cleaner cuts and a better finished joint.

Once the tails are cut, turn your attention to the pins. These are cut on the adjoining workpiece, precisely matching the layout of the tails. Here, accuracy in marking is again crucial. A small error in the pin layout will immediately show up when you try to fit the joint. The process of cutting the pins is similar to cutting the tails, but the angle is reversed. As you cut, test the fit frequently, carefully adjusting your cuts to ensure a snug, but not overly tight, fit. A properly-cut dovetail joint should fit together smoothly and firmly, but still allow the pieces to be assembled without force. A slightly loose fit is preferable to one that's too tight.

The assembly of the joint involves carefully aligning the tails into their respective pins. Apply a thin layer of high-quality wood glue to the surfaces of the tails and pins to create a strong bond. Then use clamps to firmly hold the two pieces together while the glue sets. Avoid using excessive clamping force as it can squeeze out the glue and potentially damage the joint. Allow sufficient time for the glue to cure completely before attempting to work with the assembled joint.

Several types of dovetails exist, each suited to different applications. Through dovetails are commonly used for drawers, offering strength and a pleasing appearance. Half-blind dovetails, where the tails are visible on one side and the pins on the other, are a more practical choice for certain applications, offering a neat interior. And then there are full-blind dovetails, where neither the pins nor the tails are visible, showcasing the elegance and skill of the craftsman. Selecting the appropriate type depends on your project's design requirements and the desired aesthetic outcome. Carefully consider the strengths and limitations of each to achieve optimal results.

The beauty of a well-executed dovetail joint lies not only in its strength but also in its inherent elegance. The precise geometry, the intricate interplay of perfectly fitted shapes—these all contribute to the aesthetic appeal of this traditional joinery technique. A well-crafted dovetail is a testament to the skill and precision of the woodworker, a visual delight that enriches the overall piece. It's more than just a structural element; it's an expression of craft, a detail that elevates your woodworking to a higher level. By mastering the dovetail, you're not just building a joint, you're building your legacy as a skilled craftsperson. Practicing consistently, and paying attention to the details, will allow you to truly master this cornerstone technique. Take your time, refine your technique, and celebrate the art of precise woodworking.

2.3 Box Joints

Box joints, deceptively simple in appearance, offer remarkable strength and a clean aesthetic, perfect for drawers, boxes, and even some furniture frames. Unlike the more visually striking dovetail, their strength derives from the interlocking nature of the numerous small pins and sockets. This makes them incredibly durable, resisting racking and twisting forces far better than many other joint types. The repetitive nature of the cuts, while requiring precision, is also surprisingly efficient, especially with the aid of modern tools.

Precision is key; inaccuracies will be immediately apparent. Your choice of wood also plays a vital role. Hardwoods like maple or cherry hold their shape exceptionally well and produce crisp, clean cuts, ideal for highlighting the joint's inherent beauty. Softer woods, while workable, may chip or tear more easily during cutting, potentially compromising the joint's integrity. Remember to plan your wood carefully, considering the grain direction and ensuring your stock is adequately sized for the number of pins and sockets you intend to create.

The process begins with meticulous marking. A simple jig – readily available or easily constructed – significantly aids in maintaining consistency between each cut. These jigs allow for quick and accurate marking out of the numerous tenons and mortises, ensuring a precise fit. Take your time; rushed work always shows in the final product. Accuracy in marking translates directly into a strong, flawless joint. Avoid guesswork; each cut must be carefully considered and executed with precision. This initial stage establishes the foundation for the entire joint; hence any errors here are amplified later on.

Cutting the pins and sockets requires a sharp chisel and a steady hand. Start by cutting the shoulders of your pins and sockets cleanly; these are critical for strength and appearance. It's better to make several passes with a sharp chisel, removing small amounts of material, than to attempt to remove large amounts in one fell swoop. A smooth, precise cut not only improves the joint's strength but also enhances its visual appeal. Keep your chisels sharp, and remember to frequently check your progress to ensure accuracy and avoid any unforeseen complications.

Testing the fit is crucial throughout the process. Regularly test the fit of your pins and sockets as you work. This allows you to adjust your cuts if necessary, preventing more substantial problems down the line. A snug, yet not overly tight, fit ensures both strength and ease of assembly. Too tight and the pieces won't assemble easily, too loose and the joint lacks the necessary strength. A dry-fit test is your best friend, allowing for subtle adjustments without compromising the already-made cuts.

Gluing and assembly are the final steps. A thin, even coat of quality wood glue is essential for creating a strong, lasting bond. Carefully apply the glue to the surfaces, ensuring it is spread evenly to maximize contact and prevent weak points. Then, gently tap the pieces together, ensuring proper alignment. Clamping is crucial; use appropriate clamps and apply even pressure, ensuring all surfaces are firmly held in place. Let the glue cure completely according to the manufacturer's instructions. Once it is set, remove the clamps carefully and admire your perfectly made box joint.

Box joints represent a considerable step up in woodworking skill. The consistent repetition of cuts demands patience and precision, fostering a deeper understanding of woodworking principles, not just mechanically but aesthetically. Mastering this joint opens doors to more complex projects, reinforcing confidence and skill. The clean lines and inherent strength offered by box joints are a testament to the beauty and functionality of this classic woodworking technique.

Remember to practice your cutting, and soon, the repetitive nature of creating this joint will become second nature. The result? A robust and visually appealing joint to be proud of. And, the satisfaction is substantial.

3 Advanced Techniques

Laminating wood opens up a world of possibilities. Imagine creating a tabletop far wider than any single board could provide, or crafting a gracefully curved chair back without the limitations of naturally bent wood. This involves carefully gluing together thinner pieces of wood, often using specialized adhesives and clamping techniques to ensure a strong, stable bond. Careful selection of wood, ensuring consistent grain direction across the layers, is crucial for preventing warping or cracking. The use of a surface planer after gluing is essential for achieving a flawlessly smooth and even surface. Remember to always use appropriate clamps and ensure even pressure distribution across the entire laminate. Failure to do so can result in uneven glue distribution and potential structural weaknesses. Experiment with different wood species, consider combining contrasting grains for added visual interest, or explore the world of edge-banding to finish the edges of your laminate with precision and style.

Curved woodworking is a challenge that rewards patience and precision. Steam bending is one technique, using heat and moisture to soften the wood fibers, making them more pliable for shaping around forms. This requires specialized equipment, but the results—elegant curves in furniture legs, chair backs, or decorative elements—are well worth the effort. Alternatively, consider using a wood bending jig, which provides controlled pressure and allows for bending thinner stock without the need for steaming. Remember that the selection of wood is paramount; some species are more suitable for bending than others. This technique requires a deep understanding of wood properties and their response to stress and pressure. Always be cautious and wear appropriate safety equipment during the bending process, since the steam and bending pressures can be substantial.

Inlays and veneering transform the ordinary into the extraordinary. Inlays involve embedding pieces of contrasting wood (or other materials) into a larger surface, creating intricate designs and patterns. This technique necessitates meticulous planning and precise execution. Veneering uses thin sheets of high-quality wood to cover a less expensive or more unstable substrate. This is a cost-effective way to add beauty and durability to projects. Many different tools are required, from extremely sharp chisels for precise inlay work to specialized veneer presses for creating flawlessly smooth surfaces. The patience to accurately cut and fit the pieces is crucial, ensuring a snug and invisible fit. A thorough knowledge of woodworking adhesives is fundamental to successful inlay and veneering, as the bond must be strong and seamless.

Wood finishing brings your creation to life, showcasing the beauty of the grain and protecting it from the elements. Basic finishing involves sanding to a smooth surface, followed by the application of stain, sealant, and topcoat. The order and specific products are dependent on the desired outcome and the type of wood used. Sanding is often done in multiple stages, using increasingly finer grits of sandpaper, to achieve a mirror-like surface. Understanding the properties of various stains, from water-based to oil-based, is essential for achieving the desired color and depth. Sealants protect the wood from moisture and stains, providing a barrier against damage. Finally, topcoats provide a durable protective layer and enhance the overall beauty and longevity of your finished product. Each step requires a precise approach, otherwise mistakes can ruin the final finish. Thorough cleaning and preparation are fundamental for achieving professional results.

Advanced finishing techniques elevate your work to the next level. Consider exploring French polishing, a time-honored method resulting in a deep, lustrous sheen. Or, explore the use of specialized finishes like shellac or lacquer, which offer unique properties and finishes. Understanding the properties of each finish and their suitability for specific applications is important. Careful preparation and the application of multiple thin coats, rather than one thick coat, are crucial for achieving a blemish-free and durable finish. Consider experimenting with different finishing techniques and creating unique color blends by mixing different stains or using glazes to add depth and richness to your projects. The final step is often polishing, to remove any blemishes or imperfections, revealing the wood's natural beauty. Remember, a well-executed finish is the hallmark of quality woodworking, showcasing your skill and dedication to detail.

3.1 Laminating Techniques

Laminating allows joiners to create pieces far beyond the limitations of single wood planks. Imagine crafting a tabletop wider than any single tree could provide, or shaping a complex curve impossible to achieve with solid wood. This is the power of lamination. It's a process of bonding multiple layers of wood together, typically using adhesive, creating a single, stronger, and more stable piece. The strength derived from this process allows for incredible design freedom. Think of the possibilities.

Proper preparation is key. Start by selecting your wood carefully. Choose boards of consistent thickness, free from significant defects like knots or cracks that might compromise the finished piece. The grain direction should be consistent across all layers, ideally running parallel for maximum strength. Pay close attention to the moisture content; it should be uniform to prevent warping or cracking after bonding. If using different species, ensure compatibility in terms of movement and expansion properties. Careful selection at the outset minimizes future headaches. This careful preparation will save you considerable time and frustration.

The adhesive plays a crucial role. Several adhesives exist, each with its own properties and strengths. Some are designed for faster curing times, while others offer greater strength. A well-known choice is wood glue, often used for its reliability. Ensure you have a suitable adhesive that is appropriate for your wood type and the project's demands. Remember always to follow the manufacturer's instructions meticulously regarding application, open time, and clamping pressure. Improper adhesive application is a frequent source of failure. A thin, even spread is essential for a strong bond.

Once your adhesive is chosen and your wood is prepared, it is time to assemble the laminations. Accurate alignment of each layer is paramount, for a smooth, even final piece. Use clamps to ensure even pressure across the entire surface. Apply pressure firmly but carefully, to avoid crushing the wood. If uneven pressure is applied, it can cause the layers to become uneven or even to separate. Apply the clamps evenly, and adjust them as needed to keep the pieces securely together. Consider using caul blocks to distribute the clamping pressure, preventing indentations or marking of the wood.

The clamping process requires patience. Allow sufficient curing time for the adhesive to fully bond. This time varies depending on the adhesive used and the environmental conditions. Consult your adhesive's instructions for the appropriate curing period. Premature release of the clamps can result in a weak joint or cause warping. Patience and proper adherence to the manufacturer's instructions will result in a more effective and successful process. Let the piece sit undisturbed during this crucial phase.

After the adhesive has cured completely, carefully remove the clamps. Then inspect the laminated piece. You'll likely need to perform some planing or sanding to achieve a perfectly smooth surface. This process will depend on the adhesive used and any potential glue squeeze-out. Pay attention to the edges as you finish; these are often the areas that will need the most attention. Precise finishing ensures a professional look and prevents visible seams or irregularities.

Beyond basic straight laminations, this technique opens up a world of possibilities. You can create curves, using bending techniques and specialized forms, to create stunning furniture pieces. By carefully shaping individual layers before gluing, complex curved forms can be easily achieved. The laminated structure's strength accommodates these shapes, avoiding the splitting that might occur in solid wood. Curved laminated structures are commonly seen in high-end furniture design. Beyond furniture, this technique finds application in a wide range of projects, from boat building to musical instrument construction. The versatility and strength of laminated wood make it a preferred material in many fields. Understanding the principles and mastering the technique opens doors to a variety of projects with unique possibilities. You may even find it becomes one of your favourite techniques in the future. Remember the key elements – meticulous preparation, careful adhesive application, and patient clamping. With practice and attention to detail, you'll soon be creating stunning laminated masterpieces.

3.2 Curved Woodworking

Working with curved wood presents unique challenges, demanding a different approach than straight-line joinery. Unlike straight cuts, bending wood requires careful planning and execution to avoid cracking or splitting, crucial for successful projects. Understanding the wood's grain and its natural flexibility is paramount. Different wood species have varying degrees of flexibility; some, like ash or hickory, bend more readily than others, such as oak or maple. This knowledge informs your wood selection, impacting the overall design and construction process.

The most common method for bending wood involves steaming. This softens the wood fibers, allowing them to conform to a mold or form. However, steaming requires careful temperature control to prevent burning or over-softening the wood. Too much steam can weaken the wood, while insufficient steaming leaves it too brittle to bend effectively. The duration of steaming depends on the wood's thickness and species, requiring experience and careful observation to achieve optimal results. Improper steaming is a common cause of failure in curved woodworking.

Another technique utilizes bending wood using a jig and clamps. This method employs a form around which the wood is gradually bent and secured using clamps. The use of a jig is essential for ensuring consistent curvature and avoiding irregularities. The jig should be designed according to the desired curve and the dimensions of the wood piece. It's crucial to gradually increase the bending pressure to prevent sudden breaks. This careful process demands patience and precision, as rushing the process can lead to cracking.

A less common but highly effective technique involves laminating thinner pieces of wood together to create a curved shape. This method is particularly suitable for complex curves or for creating stronger, more stable pieces. Each layer of wood should be carefully aligned and glued together, using a strong, waterproof adhesive designed for woodworking. This meticulous process requires attention to detail to ensure a seamless and robust final product, free of gaps or imperfections.

Once the wood is bent to the desired shape, it's essential to secure it in place until the glue or steam-softened wood has set properly. Clamping is vital in ensuring the curve maintains its shape while the adhesive cures or the wood dries. Proper clamping avoids warping or rebounding after the wood is released. Different types of clamps – bar clamps, hand screws, or specialized bending clamps – may be necessary depending on the project's complexity and the wood's dimensions. The correct clamping pressure must be applied without crushing or damaging the wood.

Finishing curved woodwork demands particular care. Sanding curved surfaces can be challenging, requiring the use of flexible sanding tools and various grits of sandpaper. Starting with coarser grits and gradually moving to finer grits ensures a smooth, even finish. This process requires patience and attention to detail, as any irregularities will be more noticeable on a curved surface than on a flat one. Applying finishes such as varnish or lacquer also requires a different approach, requiring multiple thin coats to avoid runs or drips.

Beyond the technical aspects, successful curved woodworking also involves careful design consideration. Understanding how the curves interact with the wood's grain and the forces applied to the finished piece is key to creating durable and aesthetically pleasing projects. The design process should incorporate the unique challenges of bending wood, which often necessitates compromises in the initial design, ensuring that the finished product is strong and reliable. It's a constant balancing act of aesthetic appeal and structural integrity. Careful consideration is paramount.

Remember, practice is essential to mastering curved woodworking. Start with simpler projects to develop the necessary skills and experience before tackling more complex curves and designs. Each project becomes a learning opportunity, allowing you to refine your techniques and understand the nuances of working with curved wood. Don't be discouraged by initial failures; they're invaluable learning experiences. Persistence and a willingness to experiment are key ingredients in mastering this demanding but highly rewarding aspect of fine woodworking. The journey is just as important as the destination. The more you work with this material, the more you will appreciate the craft.

3.3 Inlays & Veneering

Inlays and veneering elevate woodworking from functional to artistic. These techniques, while demanding precision, offer remarkable opportunities to personalize and enhance your creations. Mastering them transforms simple pieces into stunning works of art. The subtle shimmer of a carefully chosen veneer or the striking contrast of an inlaid design can completely redefine the piece's character and value, adding sophistication and visual interest. It's a rewarding skill to acquire and a testament to dedication and precision.

Veneering involves applying thin layers of attractive wood, often a more expensive or visually interesting species, to a less costly substrate. This allows you to create furniture pieces with the beauty of exotic hardwoods without the prohibitive cost. For beginners, start with pre-made veneer sheets; these are easier to handle and ensure consistent thickness. Accurately cut veneer requires a sharp blade and a steady hand. Experiment with various cutting methods – from using a utility knife to a more precise veneer saw – to see which yields the best results for your style and the material you're working with. Remember that patience and precision are key; rushing the process often leads to mistakes.

Preparing the substrate is just as vital. It should be flawlessly smooth and flat, providing a stable, even base for the veneer. Any imperfections will show through. I recommend using a high-quality wood glue, spreading it thinly and evenly to prevent bubbling and ensure complete adhesion. The application process demands a light touch; avoid applying excessive pressure, which can lead to damage or uneven adhesion. Accurate measurements, precise cuts, and a careful application will save you time and effort down the line and ensure a high-quality result.

Inlays, in contrast, involve embedding small pieces of contrasting wood or other materials into a larger piece. This creates intricate patterns and designs, adding depth and visual complexity to your work. From simple geometric patterns to complex floral designs, the possibilities are virtually limitless. Begin with simpler designs to gain confidence and precision before moving onto more elaborate projects. Precise cutting is paramount, ensuring a perfect fit to avoid gaps or misalignments. A router or carving tools can be used to create intricate designs, but the practice is essential for the most successful result.

Choosing the right wood species for both inlays and veneers is critical. Consider the color, grain, and overall aesthetic to complement the base wood. Contrasting colors often create a dramatic effect, while similar tones offer subtle elegance. The stability of the wood is also a concern, selecting hardwoods that are less prone to warping or cracking. You'll find detailed tables detailing the properties of numerous wood species in numerous woodworking guides and reference books. Don't hesitate to experiment, noting the successful combinations for future projects.

Before starting, meticulously plan your design. Create detailed sketches and templates to ensure accurate cutting and placement. Transferring these markings onto the wood accurately is crucial to a successful inlay or veneer project. Use sharp pencils and marking tools to avoid errors and ensure the project goes smoothly. There are many techniques to accurately transfer patterns onto your wood; experiment to find the method you find the most accurate.

Once the pieces are cut and the glue is applied, use clamps or weights to ensure even pressure during the drying process. Allow adequate drying time for a strong and lasting bond. After the glue has dried completely, use fine sandpaper to smooth any uneven surfaces. Be patient and work methodically to avoid sanding too aggressively, this can damage your piece. A sharp blade, a gentle touch and patience are your key allies here.

Finishing is the final, and often overlooked, step. It protects the wood and enhances its beauty. Choose finishes that complement the wood species used and bring out their natural characteristics. For example, a clear varnish or shellac may show the color, grain, and other characteristics in the clearest light. An oil finish might be a better choice if you want a more natural feel, but you must be aware that it may darken your wood. Experiment and try different finishes to understand their effect on various wood types.

Mastering inlays and veneering requires patience, practice, and attention to detail. It's a journey of refining skills, discovering new techniques, and pushing creative boundaries. Don't be discouraged by early mistakes; treat them as learning opportunities, and each completed project will refine your approach and improve your skills. The end result, however, is a testament to your craftsmanship, transforming ordinary pieces of wood into unique and beautiful works of art. The pride you take in your completed pieces will be well worth the effort. As you refine your abilities, your inlaid and veneered pieces will become prized possessions, representing hours of dedication and meticulous detail.

3.4 Wood Finishing Basics

Preparing your wood for finishing is crucial. Start by ensuring your project is completely sanded, moving through progressively finer grits of sandpaper, say from 80-grit to 220-grit or even higher, depending on the wood's hardness and the desired smoothness. This smooth surface is the foundation for a flawless finish. Pay close attention to removing any sanding marks and achieve a consistently even texture across the entire piece. Don't rush this step; a well-sanded surface will drastically improve your final result, making the application of your chosen finish much easier and the end product much more visually appealing.

Next comes cleaning. After sanding, dust and debris inevitably remain on your wood's surface. Wipe it thoroughly with a tack cloth. A tack cloth is a cheesecloth-like material that's lightly treated with a sticky substance; it's inexpensive and readily available at most hardware stores. This removes even microscopic particles that sandpapering might not have fully eliminated. Failure to do this properly could lead to imperfections in your final finish – tiny bits of dust becoming trapped under the finish, creating distracting flaws. A clean surface is paramount, so take your time with this important prep step.

Now, let's talk about the finishes themselves. There's a vast world of options: from simple water-based polyurethane to more exotic shellacs and lacquers. Water-based polyurethanes are beginner-friendly, easy to clean up with soap and water, and provide a durable, clear coat. However, they might not be as hard-wearing as oil-based polyurethanes. Oil-based polyurethanes offer superior hardness and durability. However, they require mineral spirits for cleaning which can be a little messier. Shellac, with its characteristic amber hue, offers a warm, elegant finish and is historically favored, but it's less durable and more sensitive to moisture than polyurethane. Lacquers dry quickly and offer a hard, durable finish. Researching various finishes is important, as the choice will depend greatly on your project and the desired final look.

Applying the finish is a skill developed with practice. Always follow the manufacturer's instructions on your chosen product. Generally, thin coats are better than thick ones. Multiple thin coats allow for better penetration, leveling out, and minimizing runs and drips – especially important for larger surfaces. Apply the finish evenly using a quality brush, or you could use a spray gun for a perfectly even coat, especially when working with large furniture projects. A spray gun allows you to build up a thin, even coat without brushstrokes, yielding a cleaner, professional appearance. Whether you use a brush or a spray gun, avoid heavy application; patience is key. Let each coat dry completely before applying the next – this drying time varies depending on the finish, the temperature, and the humidity. Check the manufacturer's recommended drying time between coats, but it's generally a good rule of thumb to wait at least 24 hours.

Finally, after applying your chosen finish and letting it dry completely, a light sanding with very fine-grit sandpaper (around 320-grit) can be beneficial. This will remove any minor imperfections or dust nibs that may have settled on the final coat. After this light sanding, always use a tack cloth again to remove any dust. This final step helps to create a perfectly smooth, even finish. This final step often gets overlooked but can significantly improve the overall look and feel of your finished product. A little extra care at this stage enhances both the aesthetic and tactile experience of your work. Remember to always wear appropriate safety equipment – respirators to avoid inhaling fumes, safety glasses to protect your eyes, and gloves to protect your hands. This is paramount for your health and safety. With careful preparation, selection of the right finish, meticulous application, and finishing touches, you'll enhance the beauty of your wood creation.

3.5 Advanced Finishing

French polishing, a centuries-old technique, offers a deep, lustrous sheen unmatched by modern methods. It involves building up thin layers of shellac, using a pad of linen or cotton, rubbing in a figure-eight motion. Patience is paramount; multiple coats are needed, each meticulously smoothed and polished, revealing the wood's natural beauty.

The process requires careful control of the shellac solution's consistency, achieving a balance to ensure a smooth, even surface without imperfections. This demands consistent attention to detail throughout. Achieving a truly exceptional French polish requires practice and a keen eye, but the rewards are well worth the effort. You'll find it particularly suited to finer pieces, where a high-quality finish enhances the overall aesthetic appeal.

Consider the use of oil finishes. Linseed oil, tung oil, or a blend, penetrates the wood deeply, nourishing it and bringing out the natural grain. Unlike varnish, oil finishes offer a more subtle, less protective coating. They're ideal for pieces that will see regular use and handling, as they allow the wood to "breathe," preventing cracking or warping. Applying oil finishes correctly requires a careful process, typically involving several coats rubbed in, allowed to soak, and then wiped off before drying. Never apply excess oil; less is often more when working with oil finishes to prevent sticky or uneven surfaces. This straightforward process is perfect for both experienced and beginner joiners.

Waxes add a final layer of protection and depth. Beeswax, carnauba wax, or a mixture, provides a subtle sheen and enhances the wood's natural color. Waxing is often the last step in a multi-stage finishing process, adding a smooth, protective layer that resists minor scratches and water damage. The application involves carefully rubbing the wax into the wood, ensuring even coverage, and then buffing to a high gloss.

Remember that a high-quality wax will require less effort and generally offer better protection and longer lasting finishes. It's a simple process with dramatic results, enhancing the final look of your project and protecting it from everyday wear and tear.

For a more contemporary look, consider using lacquer. Lacquer sprays provide a quick and even coat, but they demand a clean, well-ventilated workspace. This finish offers a hard, durable surface that's resistant to scratches and water damage. Be sure to always follow manufacturer's directions when applying lacquers and sprays, especially regarding application safety and drying time. Multiple thin coats applied at even intervals will produce the best final finish, minimizing the risk of drips and creating a smooth, even look. With lacquer, preparation is key; ensure a flawlessly smooth surface beforehand, as lacquer will highlight any imperfections. The even coat produced by lacquers offers a modern appeal to finished pieces.

Beyond the specific finishes, remember the importance of preparation. Before applying any finish, ensure your piece is thoroughly sanded, starting with coarser grits and moving to finer ones. This creates a smooth surface that allows the finish to adhere properly and evenly, preventing blotchiness or unevenness. This thorough sanding is crucial for any kind of finish to produce consistent, high-quality results that last. A well-sanded piece is the foundation of a stunning finish, no matter what type of wood or design is involved in your project. Take your time; rushing this step will result in a sub-par outcome.

Finally, understanding the wood's properties is crucial. Different wood species react differently to various finishes. Some woods, such as mahogany, readily accept oil finishes, while others, like cherry, may require more attention to achieve a consistent look. Experimenting with samples of the wood you're using is essential before moving on to the main project. This test run allows you to see how the finish behaves on your particular wood, giving you a chance to adjust the process as needed and perfect the application method. This prevents surprises on the final project. Remember that the unique qualities of each wood will affect how well it interacts with various finishes.

4 Furniture Projects

Let's start with the simple stool. This project is perfect for beginners, allowing you to practice your mortise and tenon joints—hopefully, you've mastered them by now. You'll need three pieces of wood: two for the legs, carefully cut to the same length, and a slightly wider piece for the seat. Remember to account for the tenons when measuring the leg lengths, ensuring a snug fit with the seat. Precise cuts are key here; a poorly cut tenon will weaken the entire stool. Take your time and double-check your measurements. Use your marking gauge to accurately transfer your measurements. A sharp chisel is your best friend in this process.

Next, we move onto the small table. This project offers a chance to showcase more advanced techniques. I recommend using a slightly more complex design, perhaps incorporating dovetail joints for the tabletop frame. This will offer a satisfying challenge, pushing your skills learned earlier in this book. Remember the importance of choosing the right wood for your tabletop: its strength is paramount in bearing the weight of objects. Choose a wood known for its durability and stability. Oak or maple would be excellent choices, providing both beauty and resilience. Once again, precise measuring and meticulous cutting will be your most valuable allies.

Before assembly, carefully sand all your pieces to create a smooth surface. This will make the final project more appealing and prevent splinters. This sanding is important, and doing it well before assembly will save you considerable time and frustration later on. Pay close attention to the edges, ensuring they are even and smooth. Any roughness will show through in the finished piece. Remember that preparation is half the battle; thorough sanding makes all the difference. Using progressively finer grits of sandpaper is important for a professional finish.

Now, let's talk about finishing. Once your table and stool are assembled, you'll want to protect your hard work. A good finish enhances the beauty of the wood and safeguards it from damage. Begin with a light sanding of the entire piece to remove any small imperfections. Then apply a wood conditioner to ensure even stain absorption. The type of finish you choose—oil, varnish, or lacquer—depends on the level of protection and sheen you desire. Consider the setting where these pieces will reside to help inform your finishing choice.

For your stool, a simple oil finish could be perfect, providing a natural look. For the table, you might prefer a more protective finish, such as varnish or lacquer, to withstand daily use. Multiple thin coats, applied carefully and allowed to dry completely between applications, will yield a far superior finish to one thick coat. Remember, patience is a virtue in woodworking. Never rush the finishing process. A rushed finish will ultimately detract from the overall quality of your work. Allow adequate drying time between each coat. This patience will be rewarded by a truly professional-looking final product.

While both projects use similar principles, this chapter has focused on expanding your skill set. The stool reinforces basic joinery, while the table introduces more complex joints and finish choices, enhancing your proficiency. Remember to always prioritize safety, using appropriate personal protective equipment (PPE) throughout the process. Never rush your work; each cut, each joint, is an opportunity to improve your technique. Pay close attention to detail in every step, and you'll learn the most from these projects. These projects aren't just about building furniture; they are about refining your technique and understanding the nuances of woodworking.

With careful planning and execution, you'll create not only functional pieces but also beautiful testaments to your growing woodworking skills. Remember to photograph your progress! It's always fulfilling to look back on your journey and see how far you've come. Enjoy the process. The satisfaction derived from creating something with your own hands is immeasurable, particularly when it's something both useful and aesthetically pleasing. Building these pieces offers invaluable experience in planning, executing, and improving your woodworking abilities. It serves as a stepping stone to further projects and deeper skill refinement. The knowledge and skills gained here will form the foundation for more ambitious woodworking endeavors in the future.

4.1 Simple Stool Construction

Three sturdy legs, each a carefully chosen piece of seasoned hardwood, approximately 18 inches long and 2 inches square, form the foundation. These legs, meticulously planed and sanded to a smooth, even surface, are the heart of this simple yet elegant stool. Their strength and stability are paramount; choose wood known for its durability, such as oak or maple, ensuring a long-lasting piece. Before assembly, ensure each leg is perfectly square, as any discrepancy will manifest in a wobbly final product. Precise measurement and preparation are key to success. Next, we craft the seat. A circular piece of wood, about 12 inches in diameter and ¾ inch thick, will serve as the top. The choice of wood for the seat is primarily aesthetic; consider a contrasting grain to highlight the design or a matching wood for a unified look. Remember to select a piece free from knots or cracks that could compromise its strength or visual appeal. Before attaching the seat, carefully smooth its edges to prevent splintering and ensure a comfortable sitting experience.

The critical joinery comes in connecting the legs to the seat. A simple but effective method uses mortise and tenon joints. Cut three precisely sized mortises – rectangular recesses – into the top of each leg, carefully positioning them to distribute weight evenly. Then, create corresponding tenons – protruding pieces – on the underside of the seat. Ensure that the tenons fit snugly within the mortises for a strong, stable connection. This process necessitates precision and patience; using a mallet and chisel carefully, avoid damaging the wood. Accurate measurements are crucial.

Once the tenons are securely seated in the mortises, consider reinforcing the joint with wood glue and small, strategically placed wood screws. This will provide additional strength and prevent any movement over time. Let the glue set completely before continuing to the next step. This step is about adding a crucial element of strength and longevity to the design. It may seem like an insignificant detail, but this reinforcement will significantly impact the stool's ability to stand up to daily use and the passage of time.

The final step involves finishing the stool. Sand all surfaces again, ensuring a flawless smoothness. Choose a finish that complements the wood and protects it from moisture and wear. Several coats of a high-quality varnish, polyurethane, or even a simple oil finish will enhance the beauty of the wood and extend the lifespan of the stool. Let each coat dry completely before applying the next. Remember to follow the manufacturer's instructions for drying times to ensure a successful application.

This detailed process highlights the importance of careful planning and execution. Selecting suitable wood, accurately measuring and cutting, and precisely creating the joints are paramount to the project's success. A seemingly simple stool, like this, reveals the beauty and strength of well-executed joinery. It stands as a testament to the skill and attention to detail that define the craft of fine woodworking. Every step – from choosing the wood to applying the finishing touches – adds to the final product's appeal and functionality. The satisfaction of creating something functional and beautiful with your own hands is an integral part of this woodworking journey. This project is an excellent opportunity to practice fundamental joinery techniques and to appreciate the beauty that lies in simplicity and precision. With careful attention to every detail, you'll be able to craft a sturdy and aesthetically pleasing stool, a piece that reflects your skill and attention to craftsmanship, showcasing the very essence of joinery's elegance. The meticulous care taken during construction will be evident in the final piece, demonstrating a commitment to quality and a respect for the materials and the craft. Remember, the end result is a testament to the process; it is not just a stool, it's a representation of your woodworking journey and a piece you can be proud of. Therefore, take pride in each step, for it all contributes to the final, polished creation.

4.2 Building a Small Table

We'll build a simple, sturdy table, perfect for a small space or as a beginner's project. You'll need some basic woodworking tools: a hand saw or a circular saw, a drill, chisels, clamps, sandpaper, and measuring tools. Remember safety first; always wear safety glasses and appropriate hearing protection. Choose your wood carefully; hardwoods like oak or maple are durable but more challenging to work with than softer woods like pine. For this project, pine would be a good choice for ease of handling.

Start by cutting your lumber to size. Accurately measuring and cutting are crucial for a successful project. Your table needs four legs, a tabletop, and two support rails for added stability. Plan the dimensions based on your desired table size. Ensure your cuts are straight and square. A miter box can greatly improve the accuracy of your cuts, especially for the legs.

Next, assemble the leg supports. For this small table, we'll use simple mortise and tenon joints. Cut precise mortises into the legs, ensuring they're perpendicular to the wood grain for maximum strength. The tenons should be accurately sized to fit snugly into the mortises. Test the fit repeatedly during the cutting process; it's easier to adjust a slightly oversized tenon than a too-small one. You'll need to dry fit the joints to check for proper alignment before gluing and clamping.

Once the leg supports are assembled, carefully glue and clamp them. Use a good quality wood glue and apply even pressure with your clamps. Let the glue cure completely according to the manufacturer's instructions. This drying time is essential for a strong and lasting bond; rushing this step will compromise the table's structural integrity.

Now, we'll add the table top. For simplicity, we'll use a single piece of wood for the top, but you could use multiple pieces joined with butt joints, reinforced with biscuits or dowels. Carefully attach the tabletop to the leg assembly. You can use screws for strength, but countersink them and fill the holes with wood filler for a cleaner look. Remember to pre-drill the holes to prevent the wood from splitting.

Finally, sand the entire table. Start with coarser sandpaper to remove any imperfections, then move to finer grades to achieve a smooth finish. Sanding is crucial to achieving a professional and elegant finish. Don't neglect the corners and edges; pay attention to those details. The final sanding should ensure the whole surface is smooth to the touch. Apply a finish of your choice, such as varnish or stain, following the manufacturer's instructions. This will protect the wood and enhance its appearance. Multiple thin coats are generally better than one thick coat for a beautiful and even finish. Apply each coat evenly, allowing it to dry properly between each application, as you would always do. This step is crucial to providing lasting protection for your creation, extending its life considerably.

Remember, patience and precision are key in woodworking. Take your time, and don't be afraid to make mistakes; they are valuable learning opportunities. Enjoy the process! This simple table project provides a great introduction to building more complex pieces. With practice, you'll improve your techniques and build more intricate furniture. Remember that the key to successful joinery is precise measurements, careful cutting, and attention to detail in every step. So take your time, enjoy the process, and let your creativity flow! You'll soon find yourself confident in your woodworking abilities. Every project, from small tables to more intricate structures, helps you refine your technique and deepen your understanding. The more you practice, the easier it will become to craft precise and elegant pieces, building confidence with every project you tackle. So keep practicing and enjoy the satisfaction of building your own beautiful furniture. It's a skill that will serve you well for years to come. Don't be afraid to experiment with different designs and woods as you gain experience. There are many other styles you can try out to create your own bespoke items. Soon enough, you'll be building more complex pieces of furniture with confidence. The joy of creating something with your own hands is a rewarding experience.

5 Cabinet Making

Cabinet construction begins with meticulous planning. Accurate measurements are paramount; even a slight error can compromise the entire project. Start by creating detailed drawings, specifying dimensions for each component – sides, top, bottom, shelves, and back. Consider the type of wood; hardwoods like oak or maple offer durability, while softer woods like pine are easier to work with but may require more care. Remember to account for wood movement due to seasonal changes; proper joinery and construction techniques mitigate this. Predrilling pilot holes is crucial for preventing wood splitting when assembling the cabinet.

Next, cut the wood components according to your plans. Use a precision saw, whether hand or power, ensuring clean, square cuts. Pay close attention to the grain direction; cutting against the grain can lead to splintering and weaker joints. Once cut, carefully sand all edges and surfaces to create a smooth finish. This step is often overlooked, but it significantly impacts the final look and feel of the cabinet. A smooth surface also makes assembly easier and safer. Consistent sanding with progressively finer grit sandpaper is key to a professional finish. Proper dust collection during sanding is vital for both health and environmental reasons.

Assembling the cabinet requires careful attention to detail. For the carcass, robust joinery is essential. Mortise and tenon joints provide excellent strength and stability, ideal for the side-to-bottom connections. Ensure the tenons fit snugly into the mortises and use wood glue for a strong bond. Clamping is crucial to ensure proper alignment and even pressure during glue drying. Accurate measurements and careful execution are critical for a seamless, visually appealing result. For the back, consider using plywood for strength, attaching it with a brad nailer or screws.

Cabinet doors and drawers are integral to functionality and aesthetics. Drawer construction usually involves dovetail joints for the sides, providing beautiful strength and a timeless appearance. Alternatively, box joints offer a strong alternative, particularly for wider drawers. Using a router jig can help increase precision and consistency when creating these joints. Precise cuts are essential here; a well-fitting drawer slides effortlessly and quietly. For the drawer fronts, consider using face-frame construction, which creates a strong and visually appealing frame around the drawer opening.

Door construction can use various methods, from simple frame-and-panel designs to more intricate raised-panel doors. The choice depends on your skill level and the desired aesthetic. Hinges should be carefully selected and installed, ensuring proper alignment and smooth operation. Consider using high-quality hinges for longevity and durability. Remember to account for appropriate spacing between doors and drawers for comfortable access and operation. Proper hinge placement is important for a well-functioning door. Pay attention to both the position and the type of hinge chosen, as various hinges suit different projects.

Finishing your cabinet is the final, yet crucial step. Begin with sanding, progressing through finer grits to ensure a smooth surface free from imperfections. Then, apply a stain or paint according to your design. Ensure proper ventilation and follow the manufacturer's instructions. If using stain, several thin coats yield a more even result compared to one thick coat. Seal the wood with a clear coat to protect the finish from damage and enhance its longevity, adding a protective barrier against moisture and scratches. Allow ample drying time between coats to avoid imperfections and ensure the finish sets properly. Choose a finish appropriate for the cabinet's intended use and location. For kitchen cabinets, consider a durable finish resistant to moisture and grease.

The final stages involve installing the hardware—knobs, pulls, and hinges. Ensure they are correctly placed and securely attached. Take your time, making sure everything functions smoothly. A well-fitted cabinet with high-quality hardware will enhance its functionality and aesthetic appeal. Always double-check the installation before considering the project complete. A final inspection will ensure all aspects align with your original plan, offering a final opportunity to correct any minor imperfections. Take pride in the craftsmanship of your creation. Your hard work and attention to detail will result in a beautiful, functional, and long-lasting cabinet.

5.1 Cabinet Construction

Cabinet construction begins with meticulous planning. Accurate measurements are paramount; a poorly planned cabinet is a frustrating project, quickly turning into a wobbly, ill-fitting eyesore. Draw detailed plans, including all dimensions and cut lists, before touching your lumber. Consider using design software for complex projects, creating three-dimensional models to visualize the final product and catch potential issues early. This is especially helpful with intricate joinery or unusual angles.

Next comes the selection of your materials. The type of wood profoundly affects both the aesthetic appeal and the structural integrity of your cabinet. Hardwoods like maple, cherry, or oak offer superior strength and durability, making them ideal for cabinets destined for frequent use. Softwoods such as pine are more readily available and cost-effective but require careful consideration, particularly regarding the joinery used and the overall design of the piece to ensure sufficient strength. The grain direction also plays a vital role; selecting boards with straight, consistent grain results in a more stable and less prone to warping or cracking finished piece. Consider the anticipated load and environmental conditions (humidity, temperature) to make the right choice.

Cutting and shaping your wood requires precision. Using a table saw or a hand saw, ensure your cuts are accurate and square. Planing, smoothing, and sanding create the perfect surface for joinery. Any inaccuracies compound over the construction process, quickly leading to a less-than-perfect final product. Remember to take your time at each step, making sure that each cut or adjustment is executed correctly. Rushing results in rework, frustration, and ultimately, a piece that lacks the quality you envisioned. Pre-drilling holes before driving screws prevents splitting, a common issue, especially in harder woods. Moreover, clamping techniques should be carefully considered; proper clamping ensures squareness and alignment between pieces.

Assembling the cabinet's carcass requires careful attention to detail. Mortise and tenon joints offer robust strength and are suitable for this application. The precision of your joinery is critical; poorly executed joints weaken the entire structure, potentially causing the cabinet to become unstable or even collapse under stress. Once joined, carefully reinforce the structure by gluing the pieces together, ensuring the glue is applied evenly and generously for a strong bond. Clamping ensures accurate alignment and minimizes the risk of misalignment. Use appropriate clamps to keep even pressure and avoid compression of the material. The cabinet's sides, top, and bottom form the basic skeleton; accuracy during construction is vital for aesthetic appeal.

Constructing the doors and drawers involves different techniques. These components usually incorporate dovetail joints, renowned for their strength and clean aesthetic. The quality of your dovetails directly impacts the functionality and longevity of these parts. Accuracy is crucial when creating dovetails; sloppy execution makes for weak joints that can fail under normal use. Furthermore, using proper tools and techniques is key. Invest in high-quality tools to ensure smooth, precise cuts; the precision of the joinery impacts both structural integrity and overall appearance. A properly constructed drawer should glide smoothly, without sticking or binding. After assembly, check for any needed adjustments and smooth any rough edges before final finishing.

Finally, finishing is the critical last stage. It protects the wood from damage, enhances its natural beauty, and completes the cabinet's aesthetic. Sanding to a smooth finish is essential before applying any stain or sealant. Choose a finish appropriate for the type of wood and the cabinet's intended use; some finishes offer superior durability, while others enhance the wood's natural grain patterns. Consider the cabinet's placement; a cabinet in a high-traffic area requires a more resilient finish than one in a less frequented location. Multiple coats, allowing sufficient drying time between applications, are recommended for optimal protection. Careful attention to every detail ensures a high-quality finish that enhances the cabinet's overall appearance, enhancing its longevity and value.

5.2 Door & Drawer Assembly

Precisely aligning doors and drawers within a cabinet demands meticulous attention to detail. Begin by ensuring your cabinet carcass is perfectly square and level; any imperfections here will be magnified in the final assembly. Use a combination square and level religiously, adjusting as needed. Don't rush this step; it's the foundation of a professional-looking piece. Remember, even a tiny discrepancy can lead to frustrating gaps or binding.

Cutting rabbets for the drawer fronts requires sharp chisels and a steady hand. Practice on scrap wood first to get the feel for the depth and angle needed. Consistency is key; uneven rabbets will result in a wobbly or misaligned drawer front. Aim for precise, clean cuts to guarantee a smooth, flush fit. Using a router with a rabbeting bit can offer more consistency but requires extra caution to avoid tear-out, especially on softer woods.

Drawer slides are crucial for smooth operation. Precise measurements are vital, and the correct type of slide must be chosen for the weight of the drawer and its contents. Undersized slides lead to instability and premature wear, whereas oversized ones might be loose and sloppy. Use your measurements twice, then again to be sure before cutting your dadoes or grooves for the drawer runners, ensuring they are perfectly parallel and aligned. A slightly oversized groove will only compound the problem.

For the doors, hinges are the unsung heroes of smooth operation. Choose hinges appropriate for the door's weight and size. Mortise hinges require precision in chiseling mortises, both on the door and the cabinet frame. The mortises must be perfectly aligned, and the hinge leaves must fit snugly but not bind. I always recommend using a hinge jig; it greatly increases accuracy and reduces frustration. It can make even a complex project easy to realize and achieve your goals.

Door alignment is just as crucial as drawer alignment. Again, perfectly square and level cabinet sides are essential for accurate door hanging. Use shims to fine-tune alignment, ensuring a consistent gap between each door and the surrounding cabinet frame. These gaps, usually 1/16th of an inch, provide even spacing and prevent binding. Adjusting hinges requires patience and finesse; small adjustments can make a big difference in the final outcome. Remember, a well-hung door shows a craftsman's care and attention to detail.

Once hinges are in place, adjust the door position with careful measurements and use of shims. Check for even gaps across the entire door edge. Any variation will be immediately noticeable. Using your knowledge of angles and mathematics ensures a level and properly functioning door, a subtle but important skill of a true joiner.

Finally, consider the use of decorative elements on both doors and drawers. A carefully chosen knob or pull can elevate the aesthetic appeal. Ensure a proper fit and alignment for the hardware. Properly sized screws and pre-drilling pilot holes to reduce the risk of splitting the wood are important. A simple detail like this can transform a functional piece into a stunning piece of craftsmanship. A well-chosen handle can enhance the piece's overall look and feel significantly. Remember, quality hardware also contributes to the longevity and usability of your work.

5.3 Finishing Cabinets

The final step in cabinetmaking, often overlooked, is finishing. It's the culmination of all your hard work, transforming a functional structure into a beautiful piece. Applying a finish protects the wood from moisture, scratches, and everyday wear and tear. More importantly, it enhances the wood's natural beauty, bringing out the grain and color, making your craftsmanship truly shine. Choosing the right finish depends on factors like the type of wood, the intended use of the cabinet, and your personal aesthetic preferences. Consider the cabinet's location – will it be in a high-humidity environment, or a dry one? A high-gloss polyurethane is perfect for kitchen cabinets needing durability and easy cleaning, while a matte oil finish might better suit a cabinet in a dry, less-used room.

Before you apply any finish, meticulous preparation is vital. Any imperfections, from scratches to dust particles, will show under the final coat. Thoroughly sand the cabinet surfaces, starting with coarser grits to remove any major imperfections, gradually progressing to finer grits for a smooth, even surface. Paying close attention to detail is crucial; you should be able to run your hand across the surface and feel nothing but smooth, flawless wood. Remember, proper sanding is the key to a beautiful finish. After sanding, wipe down the cabinet with a tack cloth to remove any dust before applying the finish.

Apply thin, even coats of finish. This is better than applying thick coats, which may run, drip, and take forever to dry. Too many coats can also lead to uneven surfaces and cracking. Each coat should be allowed to dry completely before applying the next, allowing sufficient time for the finish to cure properly. The drying time varies depending on the type of finish you've chosen and the environmental conditions. If using a spray-on finish, ensure proper ventilation and wear a respirator to avoid inhaling harmful fumes. Always follow the manufacturer's instructions precisely.

Different finishes offer various levels of protection and aesthetic qualities. Oil-based finishes penetrate the wood, nourishing it while offering a warm, natural look. Water-based finishes are easier to clean up, dry faster, and offer excellent protection. Lacquer dries quickly and creates a hard, durable surface, ideal for high-traffic areas. Polyurethane is extremely durable and waterproof, an excellent choice for kitchen cabinets, but it can be challenging to work with. You'll need to select a finish that suits your cabinet's intended use and your own personal preferences.

Once the final coat is dry and cured, you may need to lightly sand any imperfections between coats. This may involve using ultra-fine sandpaper or steel wool, carefully removing any raised grain or dust nibs. Then, carefully buff the surface to remove any imperfections and bring out the shine. This last step may involve using a polishing compound or simply a soft cloth. Take your time; rushing the process can lead to a poor final result, diminishing the quality of your hard work.

Finishing cabinets requires patience, attention to detail, and an understanding of different finishing techniques. Don't be afraid to experiment; there is no one "right" way. Embrace the learning process. With experience, you'll develop your own preferences and refine your technique, creating stunningly finished pieces that you'll be proud of. Remember, a well-finished cabinet is more than just a piece of furniture; it's a testament to your skill and dedication. The beauty of the finished product lies not just in the wood itself, but in the care and attention to detail that went into its creation, from initial design to the final, protective coat.

The final polishing is the grand finale, a moment to appreciate the transformation. You've gone from rough lumber to a sophisticated, polished piece. From the initial planning stages to meticulously crafting each joint and finally, to the elegant finish, you've completed a journey. Each step, each decision, has contributed to the final product. Take time to admire your creation; it's a testament to your skill and the hours you've dedicated. The reward is not only the finished cabinet, but the satisfaction of a job well done, a feeling of accomplishment that's uniquely satisfying in woodworking.

The completed cabinet, gleaming under the light, is a visual representation of your expertise. Take pride in your craftsmanship, for it's a tangible example of your skill and dedication. Now, with this chapter completed, you possess the knowledge and tools to finish any cabinet project with confidence, producing pieces of lasting beauty and functionality. Remember, practice makes perfect; every project will enhance your abilities, leading to better and more satisfying results. The journey of a joiner is a continuous process of learning and refining your skills, constantly striving for excellence. This is not just a profession;

it's a craft, a passion, a rewarding journey of continuous growth.

6 Specialized Joinery

Carving demands patience and precision; it's not a rush job. Start with simpler projects, perhaps a small decorative element for a box or a frame, before tackling ambitious figures. Choose softer woods like basswood or lime for your initial attempts; their ease of carving will build your confidence. Sharpen your tools regularly – this is paramount. Dull tools tear the wood, leading to uneven cuts and frustration. Employ various carving techniques, from roughing out the shape with larger gouges to refining details with smaller tools. Remember, the process is iterative; you'll likely refine your design as you work. Consider using templates or tracing patterns onto your wood to ensure accuracy, especially when tackling symmetrical designs. Always wear safety glasses and a dust mask; carving can create fine wood particles that are hazardous to your lungs. Learn to control the depth of your cuts, avoiding digging too deeply and potentially damaging your work. Practice makes perfect; the more you carve, the more adept you'll become at controlling your tools and interpreting the wood's grain.

Don't be afraid to experiment; explore different tools, techniques, and wood types to find your style.

Turning wood opens a world of possibilities, offering a different creative outlet compared to carving. It necessitates a lathe, a specialized tool that rotates the wood while you shape it. Invest in a quality lathe that fits your budget and workspace. Start with basic shapes like bowls or spindles, mastering the fundamentals before attempting complex designs. Understanding the lathe's controls is critical; practice gradually increasing speed and experimenting with different cutting tools. Safety is crucial when using a lathe; always ensure your workpiece is securely mounted and wear safety glasses. Utilize the various turning tools available to achieve different effects, from roughing gouges to detail tools for fine shaping. Use appropriate turning speeds for each tool and wood type; high speed with inappropriate tooling risks injury. Finishing is as important as turning itself; smooth the surface using abrasives and apply a finish to protect and enhance the wood's natural beauty. Many wood turners favor oil-based finishes for their durability and enhancing the wood grain. Remember, like carving, practice is key to mastering wood turning; start with simpler projects and gradually work towards more complex designs. Experiment with different wood species and explore various finishes to discover your unique style. Consider purchasing a set of specialized carving tools – a variety of gouges, chisels, and V-tools are essential to address different carving techniques and wood types. With turning, learn the proper techniques for mounting your wood securely onto the lathe headstock and tailstock, ensuring stability during the turning process. Understanding the various turning tools, like roughing gouges, spindle gouges, and skew chisels, is crucial for achieving the desired shape and finish. Explore different wood species; each offers unique properties that influence the carving and turning process. Some woods are softer and easier to carve, while others are harder and demand more strength and technique. Develop a keen eye for detail; recognizing the wood grain and learning how to work with it is essential to achieve successful results in both carving and turning. Remember, even skilled craftspeople encounter challenges and make mistakes; part of the learning process is understanding and recovering from those imperfections. Maintain your tools carefully; keeping them sharp and clean extends their life and helps maintain the integrity of your work. Don't be afraid to seek guidance from experienced woodworkers; attending workshops or joining woodworking communities provides opportunities to learn from others and receive valuable feedback. Explore various finishes suitable for carved and turned pieces. Consider experimenting with natural oils, waxes, or specialized wood finishes to protect and enhance your finished pieces. Finally, document your progress. Take pictures or videos to keep track of your work and note any insights or challenges. This will not only serve as a reference but also allows you to assess your progress and identify areas for improvement.

6.1 Carving Techniques

Carving demands patience; it's not a race. Begin with simple, manageable projects. A small, decorative bird, perhaps, or a simple leaf pattern on a small block of wood. Don't jump into complex, intricate designs right away. Mastering the basics is key before tackling ambitious undertakings. Focus on getting comfortable with your tools and the feel of the wood. It's about the journey, the process of transforming a block of wood into something beautiful, something that reflects your skill and creativity.

Choose your wood wisely. Basswood, for its soft, even grain, is a popular choice for beginners. Its relative softness allows for easier carving, minimizing frustration and allowing for focused practice on technique. However, hardwoods like cherry or maple provide a rewarding challenge for those seeking a more durable, sophisticated final product, offering a more intricate grain that adds depth and character to the finished piece. Remember to select wood that's appropriately dried to prevent cracking or warping during the carving process.

Sharpening your tools is paramount. Dull tools lead to poor cuts, frustrating progress, and potential injuries. Learn to hone your chisels and gouges regularly, keeping them razor sharp. A sharp tool requires less pressure, ensuring cleaner cuts and increased control. Experiment with various sharpening techniques to find what works best for you and your tools. Regular maintenance ensures your carving tools remain reliable partners in your creative journey. Consider investing in a good quality sharpening stone and learning the proper techniques from a more experienced carver. Proper tool care extends their lifespan and elevates the quality of your work.

Begin with basic carving techniques: Learn the difference between relief carving (carving into the surface) and in-the-round carving (carving a three-dimensional form). Practice basic cuts like V-cuts, U-cuts, and straight cuts. Mastering these foundational cuts will enable you to create various shapes and textures. Understanding these principles allows for greater freedom in your designs and ensures a smoother, more efficient workflow, enhancing your overall experience. The beauty of carving lies in the gradual revelation of form, layer by layer, cut by cut.

Transferring your design onto the wood is crucial. You can use pencils, tracing paper, or even a projector to transfer patterns. Whatever method you choose, ensure accuracy; a poorly transferred design will affect the final outcome. Take your time; precision in this initial stage makes a significant difference in the carving process, leading to a cleaner, more refined final product. Consider using a light-colored wood for easier visibility of your design guidelines.

Carving is not only about the tools; it's also about the wood itself. Understand the wood's grain. Learn to carve with, not against, the grain. Carving against the grain can lead to splintering and uneven results. By following the grain, you'll maintain better control and achieve a much cleaner cut. This understanding enhances your ability to craft clean lines and maintain the integrity of your design. Respecting the natural properties of the wood ensures a successful, beautiful carving. Think about composition. Even simple carvings benefit from careful consideration of balance and proportion. A well-composed piece, no matter how small, is more visually appealing and engaging than a poorly planned one. Simple exercises in understanding visual weight and the negative space surrounding your carvings will significantly elevate your skills over time. Study the works of master carvers for inspiration and insights into effective composition.

Safety should always be your priority. Always wear safety glasses to protect your eyes from flying wood chips. Use a carving bench or a stable work surface to avoid accidents. Work at a comfortable pace, and take breaks when needed to avoid fatigue. Preventing injuries ensures you can continue to enjoy your passion. A safe workspace is a productive workspace. Remember, carving can be a meditative and enjoyable process when conducted safely.

As you progress, explore different carving styles and techniques. Try relief carving, where the design is carved into the surface of the wood, or carving in the round, where you sculpt a three-dimensional form. Experiment with various tools and techniques to expand your repertoire. Explore different wood species to experience their unique properties. This continuous learning is crucial to evolving your skills and creativity. The possibilities in carving are vast and offer endless opportunities for artistic expression.

Remember to start small. Mastering the fundamentals is key before tackling larger or more complex projects. The process of carving is about developing skill, understanding the wood and your tools, and enjoying the creative process. Celebrate your progress, learn from your mistakes, and most importantly, have fun. Carving, while demanding precision and patience, is ultimately a deeply rewarding and creative pursuit. The satisfaction of transforming a block of wood into a piece of art is an unparalleled experience.

6.2 Turning Wood

The lathe, a seemingly simple machine, opens a world of possibilities. Imagine the smooth, cool weight of seasoned wood in your hands, the satisfying hum of the motor, and the gradual transformation of raw material into elegant curves. Turning wood is more than just crafting; it's a dance between precision and creativity, a dialogue between the wood's grain and your vision. It begins with careful stock selection;

a perfectly straight piece is a crucial starting point. Avoid knots and cracks, particularly in areas experiencing high stress.

Selecting the right tools is essential; a well-maintained lathe, sharp chisels, and a steady hand make all the difference. Understanding the different types of chisels—roughing gouges, spindle gouges, parting tools, and skew chisels—is crucial for shaping the wood efficiently and safely. Start slowly, focusing on safety precautions. Always wear safety glasses, and consider a dust mask depending on the wood species. Begin with practice pieces of inexpensive softwood, such as pine. Mastering the basic techniques of speed control and tool manipulation is crucial before moving on to more challenging projects.

Turning a simple spindle is a fantastic starting point, building your confidence and allowing you to refine your technique. Focus on creating a consistent diameter, learning to control the depth of cut with the gouges. Achieving a smooth, uniform surface takes patience and practice; don't rush the process. A smooth finish can be achieved through various sanding techniques, using progressively finer grits of sandpaper. Always use the appropriate speed for each process. Too much speed might end up in a dangerous situation. The final polishing can be done with a specialized wood polish, achieving a glossy sheen that enhances the wood's natural beauty.

Once comfortable with spindles, you can move on to more complex shapes. Consider bowl turning, a technique requiring different tool manipulation. Hollowing out a bowl demands careful attention to the wood's grain and the appropriate tool placement to avoid damage. Understanding how the wood fibers behave under pressure is crucial for preventing tear-outs and ensuring a clean, smooth interior. The transition from roughing to finishing requires precise movements; gentle cuts create a consistent profile. Practice maintaining consistent speed and pressure to create even surfaces and prevent damage to your piece. Beyond the basic forms, the possibilities are truly endless. Creating beads, coves, and other decorative elements adds artistry and sophistication to turned projects. Experiment with different wood species; the rich variations in grain and color enhance the visual appeal. Consider using multiple types of wood in a single piece for unique designs; these contrasting textures might even add a visual element to your creation. Remember that working with hardwood necessitates a more cautious and slower approach because hardwoods might be tougher than softwoods. Safety should always be the first priority.

The art of wood turning is about more than just creating functional objects. It's about understanding the wood's properties, honing your skills, and letting your creativity flow. It's about transforming a simple piece of wood into a work of art—a testament to your skill, patience, and passion for woodworking. With practice, your pieces will become more refined, your techniques more fluid, and your ability to transform wood into captivating objects will grow exponentially. Enjoy this journey of discovery. Remember, each piece holds within it the unique potential for beauty and lasting value, a story waiting to be told. The smell of freshly turned wood, the feel of smooth, perfectly shaped curves – these are the rewards of this skilled craft.

7 Restoration & Repair

Damaged wood isn't necessarily doomed. Understanding the extent of the damage is paramount; a thorough inspection, often requiring magnification for subtle cracks, is crucial before even considering repair. Look closely for loose joints, splintering, insect infestation—even seemingly minor issues can escalate if ignored. Photographing the damage provides a record for tracking progress and a visual guide during the repair process. Remember, patience and meticulous observation are key. The process often begins with stabilization. For severely cracked pieces, injecting wood glue into the fissures, clamping them firmly, and allowing ample drying time can make all the difference. This process may require additional support using dowels or small wooden blocks to reinforce the glue line. In cases of significant loss of material, carefully matched wood patches, secured with a combination of glue and small screws or brads, can expertly restore the integrity. Never underestimate the power of a well-placed patch; done correctly, it becomes practically invisible.

Repairing joints demands a delicate touch. For loose mortise and tenon joints, reinforcing them requires careful consideration. Often, simply regluing isn't sufficient. Adding small wooden wedges into the mortise, carefully tapping them into place to tighten the joint, provides exceptional strength. For more severe damage, replacing the tenon entirely might be necessary, demanding precision cutting and careful fitting to ensure a seamless repair. The goal is to restore the original integrity of the joint, not just mask the problem.

Restoring the wood's surface requires specific techniques depending on the type of damage. For minor scratches and blemishes, careful sanding with progressively finer grits, followed by application of a suitable wood filler, can erase imperfections. For more significant damage, such as gouges or deep scratches, wood putty expertly matched to the wood's color provides a surprisingly effective solution. Remember to allow the filler ample drying time before further sanding and finishing. Patience is key; rushing this process will only lead to visible imperfections.

Refinishing damaged areas requires a methodical approach. Start by cleaning the area thoroughly with a suitable solvent to remove any dirt or grime. Then, carefully sand the surrounding area to ensure a smooth transition. Apply several thin coats of appropriate finish, allowing ample drying time between each coat. For significant damage, the entire piece may need refinishing for a uniform look. This is where your knowledge of various finishes and their application truly shines.

Your skill will ultimately determine the final result.

Preserving your finished restoration relies on understanding the enemy: environmental factors. Extreme temperature fluctuations, direct sunlight, and high humidity can all contribute to wood degradation. Storing the piece in a stable environment, away from direct sunlight and heat sources, helps protect your hard work. Regular inspections for any signs of deterioration are crucial; catch problems early, and prevent them from escalating. Employing a quality protective finish is vital; it creates a barrier, preventing damage from moisture and preventing the wood from drying out. Your investment in preservation will extend the life of the piece. The work of restoration is not merely about repair, it's about protecting a piece of history.

7.1 Assessing Damage

Carefully examine the piece, starting with a broad visual assessment. Note any obvious cracks, splits, chips, or missing sections. Pay close attention to the overall structure; is it wobbly? Are there any areas showing significant sagging or distortion? Don't rush this initial observation; a thorough initial assessment will save you time and effort later. Remember, patience is key to successful restoration. Take photos from multiple angles to document the damage; this will serve as an invaluable reference throughout the repair process. This detailed record will help you track progress and makes it much easier to understand the repair strategy.

Next, investigate the joinery. Are any joints loose, broken, or completely detached? Examine the tenons and mortises, dovetails, or any other joining method for signs of weakness or damage. Gently test the stability of each joint; wiggling them subtly can reveal hidden problems. If a joint is severely damaged, replacement might be necessary, but less invasive repairs might be sufficient for less critical damage. Prioritize your repair strategy from most important to least important.

Now, focus on the wood itself. Look for signs of rot, decay, insect infestation, or water damage. Often, these issues are hidden below the surface, meaning more careful examination is crucial. A small probe or even a thin screwdriver can gently explore the affected area to assess the extent of the damage. If rot is present, you might need to carefully remove the damaged wood and replace it with sound material. Remember that safety is paramount, so wear appropriate protective gear.

Consider the overall context of the piece. Is it antique? Does it have any historical or sentimental value? This understanding informs your repair strategy. For antique furniture, preservation might be favored over complete restoration, maintaining its original character. This historical consideration informs how you choose materials and techniques. A modern replacement on a historical piece is not always ideal, whereas a modern piece might suit modern replacements more easily. Once you've thoroughly documented all the damage – cracks, loose joints, rot, etc. – you can start creating a repair plan. Prioritize repairs based on structural integrity, considering which problems need addressing first to stabilize the piece. This might involve temporarily reinforcing weakened areas before proceeding with more complex repairs. Sketch a plan, noting the type of repair for each damaged area. Detailed sketches, alongside the photos, will serve as a useful guide. Finally, gather the necessary materials and tools. This step should be carefully planned to ensure you have everything needed before you begin, preventing disruptions and delays during the repair process. Choose your materials carefully, considering the type of wood, the adhesives, and any finishing materials needed to match the original as closely as possible. This preparation is a sign of respect for the project, and will result in a better overall result. Don't be afraid to seek advice from experienced woodworkers; an additional expert perspective can prove invaluable, especially for complex repairs.

7.2 Repairing Joints

Damaged joints are a common problem in older furniture, and their repair requires careful attention to detail. First, thoroughly assess the damage. Determine the type of joint—mortise and tenon, dovetail, or another—and the extent of the break. Examine the wood for rot or insect damage, which may complicate the repair. If significant decay is present, addressing that should take priority before attempting joint repair; otherwise, the repaired joint might fail prematurely. This initial evaluation is crucial for determining the best repair strategy.

Sometimes, a simple glue repair will suffice. Clean the broken surfaces thoroughly, removing any dust or debris. Apply a high-quality wood glue sparingly, ensuring complete contact between the broken surfaces. Clamp the joint firmly, using appropriate clamping pressure and ensuring even distribution to avoid warping. Allow the glue to cure completely, following the manufacturer's instructions. This straightforward approach is ideal for minor breaks in strong joints that haven't suffered significant damage.

For more extensive damage, particularly in older or valuable pieces, more involved repair might be needed. You might need to create new tenons or dovetails, carefully matching the existing joinery. This requires precise woodworking skills and the use of specialized tools such as chisels, saws, and possibly a router. Remember that accuracy is paramount; any discrepancy in the new pieces will weaken the joint and compromise the repair's stability. Take your time, work meticulously, and don't rush.

If the wood itself is damaged beyond simple repair, consider using wood filler or epoxy to rebuild the missing sections. Choose a filler that closely matches the wood's color and texture. Apply the filler according to the manufacturer's instructions, smoothing it carefully to avoid noticeable patching. Once the filler is fully cured, carefully shape it to match the surrounding wood. This is where patience truly pays off—take your time to achieve a natural, seamless blend, and never skimp on surface preparation; you may need several coats of filler for the best results.

After repairing the joint, the next step involves strengthening the repaired area. This can be done by applying wood glue to the entire joint, adding dowels for extra support, or even reinforcing with small, carefully placed wooden blocks. Ensure all components are well-aligned before the glue sets. The goal here is not just to reattach the broken pieces but to create a joint stronger and more durable than it was before it broke. This preventative measure will increase its longevity, improving the integrity of your repair.

Finally, the last stage of the repair process is finishing. This may involve sanding the repaired area smooth, applying a wood stain to match the surrounding wood, and finishing with a protective topcoat. Your approach to finishing should depend on the type of existing finish on the piece you are repairing. Remember to match it as closely as possible to avoid creating a visible patch. A high-quality finish protects the repaired joint from future damage and enhances the overall appearance of the piece.

Remember to always test any repair technique on scrap wood first. This allows you to practice and refine your skills before attempting the repair on your valuable piece. It also lets you verify the effectiveness of the chosen methods and materials. This precaution helps you avoid costly mistakes and ensures that you gain the most experience possible before tackling a complex repair.

Successful joint repair relies on a careful combination of assessment, appropriate techniques, and meticulous craftsmanship. By combining these elements, you can breathe new life into damaged furniture, preserving cherished pieces for many years to come. The reward of saving a well-loved piece through your craftsmanship is immeasurable; it's one of the most satisfying parts of woodworking.

7.3 Wood Restoration

Damaged wood isn't necessarily ruined wood; often, with the right approach, you can bring it back to its former glory. Understanding the type of damage is paramount. Is it a simple scratch, a deep gouge, water damage, or insect infestation? Each requires a distinct treatment. A superficial scratch might only need some careful sanding and refinishing, while a deep gouge may require wood filler and meticulous matching of grain and color. Water damage can lead to rot, needing more extensive repairs possibly involving section replacement. Insect infestation necessitates aggressive treatment to eliminate the problem and prevent further damage, often followed by thorough cleaning and treatment of the affected area.

Assessing the damage accurately requires careful observation and often the use of various tools. A magnifying glass can reveal the extent of small cracks or insect damage. A sharp probe can test the solidity of the wood, revealing areas that are soft or rotten. Understanding the wood's species is also vital as some woods are more susceptible to certain types of damage than others. For example, hardwoods tend to be more durable than softwoods, making them more resistant to damage from impacts and scratches.

Once you have a thorough understanding of the damage, you can begin the restoration process. Repairing cracks involves cleaning the area and applying wood glue, often followed by clamping to hold the pieces together until the glue sets. For larger gaps, you might need wood filler to match the color and grain of the surrounding wood as closely as possible, a task requiring patience and skill to achieve an invisible repair. Remember to sand the filler smoothly to eliminate any visible imperfections.

Dealing with water damage is more complicated and sometimes requires replacing severely affected sections. Start by removing the affected area; rotten wood needs to be cut away carefully until you reach sound material, creating a clean surface for repair. This may involve careful chisel work to achieve a clean and precise cut, avoiding unnecessary damage to the surrounding wood. You then need to replace the damaged section with a piece of wood that closely matches the original in both species and color, ensuring a seamless repair. A successful repair requires attention to detail to ensure the new section blends flawlessly with the existing wood.

Restoring the surface of the wood typically involves sanding, which should be done carefully and gradually, using progressively finer grits of sandpaper to avoid creating additional scratches or damage. After sanding, the wood might need to be cleaned and prepared for its new finish; applying a stain can enhance the wood's natural color or provide a darker shade. Multiple coats may be required for consistent color and coverage. A topcoat, such as polyurethane or varnish, protects the wood from future damage and enhances its appearance, offering long-lasting protection.

The key to successful wood restoration isn't simply fixing the damage, but blending the repair seamlessly into the existing piece. This may require several techniques and multiple attempts to match the original. If a small area is affected, the repair can be less noticeable, requiring only a small amount of careful work. A large area of damage, however, might need more extensive repair; the entire piece may need to be sanded and refinished to maintain consistency. In such cases, matching stain and topcoat is crucial.

Remember always to take your time and work carefully. Restoration is a delicate process, and rushing can result in further damage. Proper preparation and attention to detail are key components of a successful restoration. Learning the properties of different woods and their susceptibility to damage can also make the restoration process easier, making your work more accurate and efficient. Finally, a good understanding of different wood finishes can help you choose the right one for your project, ensuring the best possible results for your restoration work. Always prioritize safety, wearing appropriate protective gear like safety glasses and dust masks. With patience, skill, and the right techniques, even severely damaged wood can be beautifully restored.

7.4 Finishing Repairs

The final sanding is crucial; don't rush it. Use progressively finer grits, moving from coarse to extra-fine, always sanding in the direction of the wood grain. This prevents scratches and ensures a smooth, even surface ready for the finish. Remember, patience is key here; a rushed sanding job will show in the final product, resulting in an uneven surface that will be difficult to correct. Careful sanding minimizes the need for excessive filling and patching later on.

After sanding, carefully examine the repaired area for any imperfections. Small gouges or scratches may still be visible, even after meticulous sanding. Use wood filler sparingly to address these, ensuring it matches the surrounding wood color as closely as possible. Allow the filler to dry completely before sanding it flush with the surface. Avoid overfilling; it's easier to add more filler than to remove excess.

For larger repairs, where significant wood loss occurred, consider using wood patches. Carefully select a piece of wood that matches the existing wood in color and grain pattern. Create a precisely fitting patch, using your joinery skills, ensuring a seamless integration with the repaired area. Apply a strong adhesive, securing the patch firmly in place. Once the adhesive cures, sand the patch flush with the surrounding surface.

Once all repairs are complete and the surface is impeccably smooth, the final finishing process begins. This isn't just about aesthetics; it's about protecting your repaired wood from damage. Begin with a suitable wood conditioner. Conditioners penetrate the wood, preparing it to accept the finish more evenly. This step is essential, particularly in the case of repairs where the wood may be less absorbent than the surrounding areas.

The choice of finish depends on the piece's intended use and desired aesthetic. A durable varnish offers excellent protection against moisture and wear, ideal for furniture or items subject to regular handling. Oil finishes, on the other hand, offer a more natural look and feel, enhancing the wood's grain. Both types require several thin coats, allowing each to dry thoroughly between applications. Avoid thick coats; they can run or cause uneven drying.

The application method also matters. For a smooth, even finish, consider using a quality brush, ensuring even coverage and avoiding brush strokes. You might also consider using a spray applicator, which delivers a thinner, more consistent coat. Regardless of the method you choose, work in a well-ventilated area, to avoid inhaling harmful fumes. This is essential regardless of whether you're using varnish or oil-based finishes. Always remember to read the product's safety instructions carefully.

Finally, after the final coat has dried completely, carefully inspect the finished repair. Are you satisfied with the result? Does it blend seamlessly with the original piece? If not, there may still be slight imperfections that can be corrected with further sanding or additional coats of finish. The time you invest now will determine the longevity and beauty of the repaired item.

Remember, repairing old wood can be a rewarding experience. You're not merely fixing damage, you're preserving a piece of history, a handcrafted item that might otherwise be lost to time or neglect. It's about more than just the mechanics; it's about preserving stories, memories and craftsmanship. By following these steps, you can return a piece of damaged wood to its former glory, extending its life and adding to its value and history. And always remember that practice makes perfect, each repair provides valuable learning experiences that will inform your approach to future projects. These repairs, once completed correctly, become a testament to your skill, your understanding of wood, and your dedication to preserving the craft of woodworking. The pride in seeing the revitalized piece will be more than worth the time and effort invested.

7.5 Preservation Techniques

The longevity of your woodworking projects hinges significantly on proper preservation. Understanding and implementing effective preservation techniques isn't just about prolonging the life of your creations; it's about safeguarding the time, effort, and artistry invested in each piece. Neglecting preservation can lead to irreversible damage, from minor surface blemishes to catastrophic structural failure. This is especially crucial for heirloom-quality pieces destined to be passed down through generations.

Protecting your finished work begins even before the final coat of varnish is applied. The selection of appropriate wood plays a vital role. Hardwoods generally possess greater inherent durability than softwoods, offering superior resistance to wear and tear, insect infestation, and environmental damage. Understanding the characteristics of various wood species, their susceptibility to moisture, and their inherent resistance to decay is crucial for making informed choices. Properly seasoned wood, with minimal moisture content, is fundamental to preventing warping, cracking, and other structural issues. This careful selection and preparation form the bedrock of long-term preservation.

Beyond wood selection, the environment in which your pieces reside significantly impacts their lifespan. Excessive humidity fosters the growth of mold and mildew, while extremely dry conditions can lead to cracking and splitting. Maintaining a stable environment, with controlled temperature and humidity, is paramount. For particularly valuable pieces, climate-controlled storage might be necessary. This isn't always practical, but even simple measures, such as avoiding direct sunlight and placing furniture away from heat sources or damp areas, can make a significant difference.

The application of appropriate finishes is another key element of preservation. Finishes provide a protective barrier against moisture, UV radiation, and insect infestation. Choosing the right finish depends on both the type of wood and the intended use of the piece. Oil-based finishes offer excellent protection, penetrating the wood and enhancing its natural beauty. However, they generally require more frequent reapplication compared to polyurethane or varnish. Polyurethane and varnish offer a durable, waterproof barrier, but they can sometimes obscure the wood's natural grain. Thorough research and experimentation are encouraged to find the perfect balance between protection and aesthetic appeal.

Furthermore, regular maintenance plays a critical role in extending the lifespan of your woodworking. This includes periodic cleaning to remove dust and debris, which can accumulate and harbor pests. Addressing minor scratches and abrasions promptly prevents larger problems from developing. While this process might seem minor, these simple acts of care significantly impact the long-term health of your creation. This preventative approach requires consistent vigilance, but it's invaluable for protecting your investment and the legacy of your work. Remember, careful maintenance is an investment in the lasting value of your craftsmanship.

Beyond the basic protective measures, consider the specific threats to your woodworking. Insect infestations, for instance, can cause irreparable damage. Regular inspections are necessary, especially in areas prone to pest activity. The use of approved insecticides should be employed only as a last resort, with careful consideration for the environment and safety. In some cases, preventative measures, such as using insect-resistant wood species or incorporating essential oils, can offer a more natural form of pest control. This responsible approach to pest management safeguards both your creations and the environment.

Finally, understanding the signs of damage and implementing timely repairs is crucial. Cracks, splits, and loose joints should be addressed promptly to prevent further deterioration. Repair techniques vary depending on the type and extent of the damage; professional restoration might be necessary for significant problems. A careful inspection process and prompt remediation will prevent small problems from escalating into significant issues. This commitment to timely repair safeguards your woodworking investments and preserves your legacy. The preservation of your creations extends beyond simple upkeep—it reflects a profound respect for your craft and the enduring value of your work.

www.ingramcontent.com/pod-product-compliance
Lightning Source LLC
LaVergne TN
LVHW022014200926
848481LV00007B/2382
* 9 7 9 8 3 4 8 5 6 5 6 4 0 *